Thanks to Dr. Aster and the magic of the comic book, I've been able to realize every physicist's dream—to get up close and personal with a black hole.

Suddenly, the night sky began to change. Einstein was right after all! Everything began to spin, and I felt like I was on a carousel going faster and faster. How beautiful it was! But something was pulling on my feet... oh yeah, I'd forgotten that when you're this close to the horizon of a black hole, [illegible] very short distances—like the length of my body.

Whoooooaah!

Look at me, I'm just like a spaghet[illegible]

When I woke up, I jumped right back into this awesome comic book. Herji and Jérémie Francfort take us through all the amazing cosmological discoveries of our time.

The 20th century was a gold mine in terms of major discoveries: the theory of relativity, the realization that the Universe is still expanding, the discovery of vestigial light from the dawn of the Universe, giant black holes at the centre of our galaxies, not to mention gravitational waves from events far away and long ago.

BIG BANGS AND BLACK HOLES

A GRAPHIC NOVEL GUIDE TO THE UNIVERSE

Every civilization on Earth has pondered the same question—how and when did the world begin? All these major discoveries are part of contemporary culture.

But what wondrous things remain for future generations to discover? What exactly is dark matter?
Why is the Universe expanding faster and faster?
What physical events were happening during the first few moments of the Big Bang?

And so on and so on...

Michel Mayor

HERJI FRANCFORT

BIG BANGS AND BLACK HOLES

A GRAPHIC NOVEL GUIDE TO THE UNIVERSE

HELVETIQ publishing has been supported by the Swiss Federal Office of Culture
with a structural grant for the years 2021-2025.

With support from:

Schweizerische Eidgenossenschaft
Confédération suisse
Confederazione Svizzera
Confederaziun svizra

Swiss Confederation

Federal Department of Home Affairs FDHA
Federal Office of Culture FOC

REPUBLIQUE
ET CANTON
DE GENEVE

Big Bangs and Black Holes
A Graphic Novel Guide to the Universe

Herji and Jérémie Francfort

Script, texts and illustrations: Herji
Theoretical background: Jérémie Francfort
Typsetting and layout: Ajša Zdravković
Translation from French: Jeffrey K. Butt
Editors: Angela Wade, Aude Pidoux, Michel Mayor, Richard Harvell
Proofreader: Ashley Curtis

ISBN: 978-3-907293-75-1
First Edition: 2023
Printed in China

Avenue des Acacias 7
CH-1006 Lausanne
Switzerland

SCRIPT AND ILLUSTRATIONS
HERJI

THEORETICAL BACKGROUND
FRANCFORT

BIG BANGS AND BLACK HOLES

A GRAPHIC NOVEL GUIDE TO THE UNIVERSE

TRANSLATED BY
JEFFREY K. BUTT

WITH THE PARTICIPATION OF
NOBEL PRIZE LAUREATE MICHEL MAYOR

A List of Things and People You Will Read About in this Book (And the Page on Which They First Appear)

* For definitions of the scientific terms used, see the glossary on p.63.

You've surely heard of the Big Bang.
People tend to think it was a huge explosion.
But that's not the case! We scientists think it was the moment the Universe ... just appeared.
SNAP!
It's hard to imagine, but that's how we see it right now.
We went from **NOTHING** ...
... to a Universe 1.0 resembling a particle soup.
Those particles were traveling at very high speeds, banging into each other.
"And then what?" you might ask.
Well, since then, a lot of things have happened in our beloved Universe.
I'd need light years to explain it all but, put simply, the Universe began to **expand**.
Keep that idea in mind–it's a very important one that'll come up again and again!

Have you seen this image before? It's what we call CMB, or cosmic microwave background.
CMB is sort of "the light" that came from the particle soup as it began to cool.
So why is that relevant?
You see, studying that cosmic light can tell us a great deal about the origins of our Universe, how it works, and how it has changed.
Speaking of observations, you might have heard of a recent astronomical discovery–gravitational waves.
No, not those!
These are vibrations that allow us to understand the very important messages coming from the farthest reaches of the Universe.
These gravitational waves owe their existence to **THE** rock star of cosmology ...
Images: ESA, NASA

BLACK HOLES!
You may remember that back in 2019 we were finally able to take a picture of one!
I know, I know, I've been bombarding you with information, so let's take a deep breath.
IIIIINNN
OOOOOOUT
So, if you want to know all about these black holes I've been going on about ...
Or where these incredible gravitational waves come from ...
If you can't wait to uncover the truth behind the famous Big Bang ...
Then fasten your seatbelts and join me on a crazy ride as we delve into the biggest mysteries of the Universe!

Ah, sorry, we'll have to save rockets for another time!
We can't dive too deep into these very important, totally amazing topics without first talking about **physics** ...
... in other words, all the "ground rules" that determine how objects move, change, and interact with each other.
And the rules we'll focus on here are the laws of **gravity and motion**.
CHAPTER 1
GRAVITY
from GALILEO to EINSTEIN
These laws are particularly relevant when it comes to very large, very heavy objects like planets, stars and galaxies.
Galileo was among the first to study gravity, even though ...
... at the time, it wasn't called that yet.
UNIVERSALIS
?
What's all the commotion outside?

Phew, nothing to worry about!
Somewhere in there is my colleague, Dr. Michel Mayor. He's the one who discovered the very first exoplanet.
MICHEEEL!
MICHEL!
MICHEL!!
Since winning the Nobel Prize in 2019, he's become a real star!
Hi Dr. Mayor. Still in one piece?
OH! Dr. Aster! Save me!
MICHMIIICH!
Quick, to my lab!
You're a real hit!
BOOM!
BANG!
BOOM!
It's REALLY gotten out of hand! There are even people who want to make a comic book with me now!
I get the sense your fans aren't going anywhere anytime soon ... I was about to visit Galileo. Wanna join me?
Galileo?
BANG!
BANG!
BANG!
BOOM!
Indeed. I was just about to tell them about the laws of gravity and motion ...
Oh, cool, hi everyone!
I'd be happy to!
But I'm done with autographs!

So here we are in Pisa in the early 17th century.
At the time, Galileo was studying two types of phenomena:
1
How objects fall to the Earth's surface.
It all started when he noticed that objects fall to the ground in a straight line.
What's most disturbing about that is that objects fall at the same speed, even when their masses are very different.*
*As long as gravity is the only force applied.
Cluck
2
How planets and the like move in the sky.
They're going every which way!
For instance, the Moon spinning around the Earth, or the planets around the Sun.
The main thing to remember is that, for Galileo, these phenomena were two separate "worlds"...
And therefore governed by different laws.
Galileo also discovered that the laws of physics apply no differently in a lab experiment than ...
... to use a modern example, in a train traveling at a constant speed. The result will always be the same!
In physics, that's what we call the "Galilean invariance."

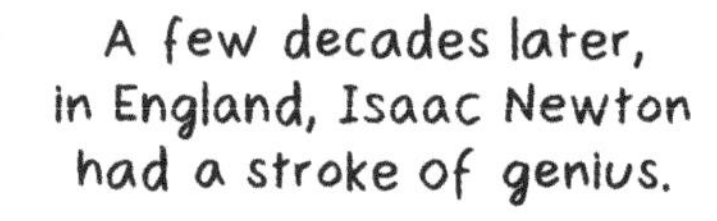

EUREKA!

He was able to show that Galileo's two "worlds" were actually one and the same!

Today, this is the field we call **gravitation.**

According to Newton, any two objects will mutually attract one another. The strength of their attractive force is more or less proportional to their respective masses and the distance between them.

Very accelerated

Not so accelerated

To describe this dynamic, he developed what are known as the three laws of "motion":

1. LAW OF INERTIA

An object at rest will remain at rest if there is no force applied to it.

If the object is set in motion and no new force is applied, its velocity will remain constant.

ZZZ

2. LAW OF FORCE AND ACCELERATION

An object to which a force is applied will accelerate in the direction of that force. Its acceleration is inversely proportionate to its mass, and vice versa.

3. LAW OF ACTION AND REACTION

"Action and reaction" means that if object A exerts a force on object B, then object B will exert an equal and opposite force on object A.

A

B

Here, the apple is object A bouncing off a head, object B.

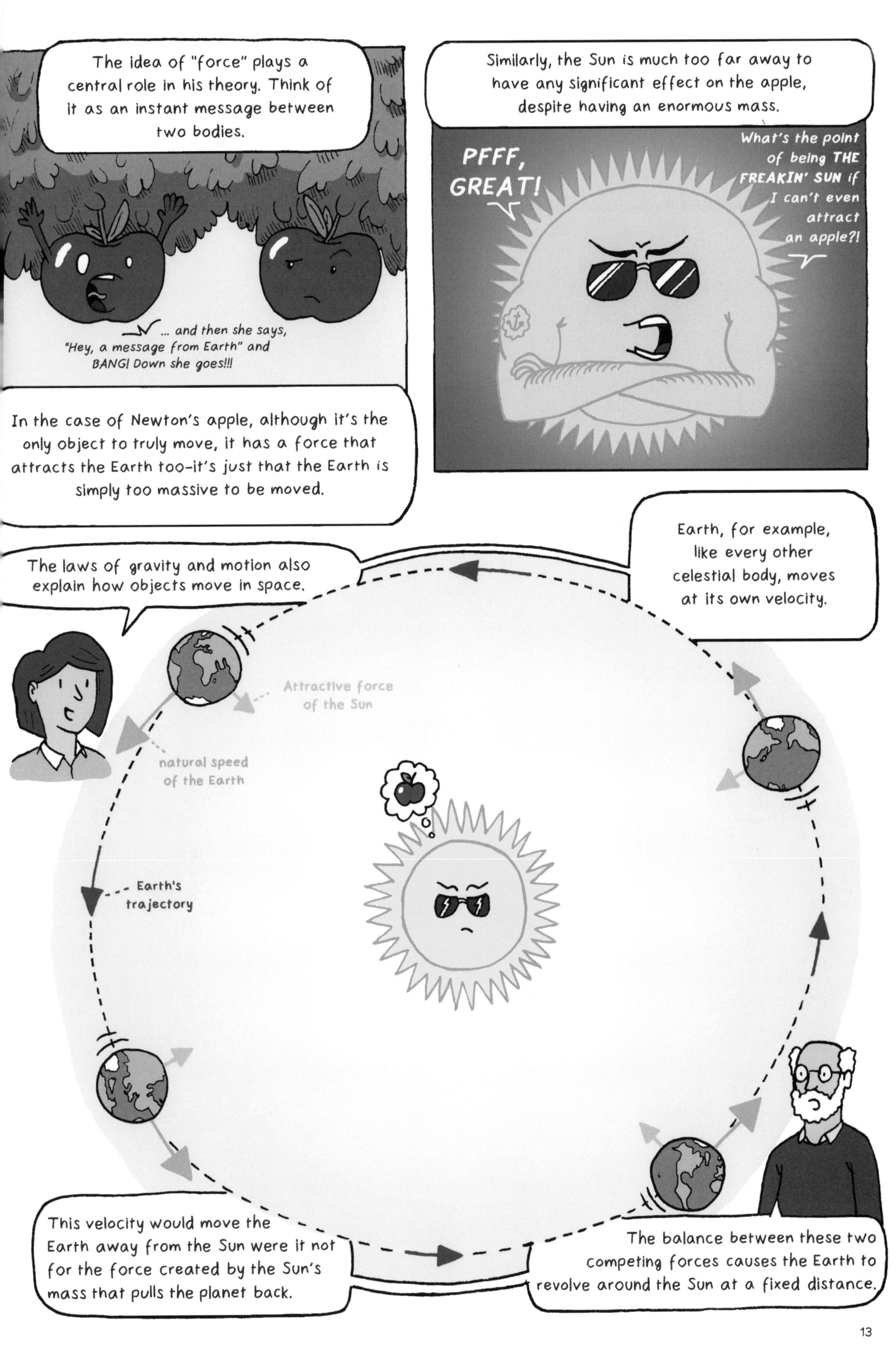
The idea of "force" plays a central role in his theory. Think of it as an instant message between two bodies.
... and then she says, "Hey, a message from Earth" and BANG! Down she goes!!!
In the case of Newton's apple, although it's the only object to truly move, it has a force that attracts the Earth too–it's just that the Earth is simply too massive to be moved.
Similarly, the Sun is much too far away to have any significant effect on the apple, despite having an enormous mass.
PFFF, GREAT!
What's the point of being THE FREAKIN' SUN if I can't even attract an apple?!
The laws of gravity and motion also explain how objects move in space.
Earth, for example, like every other celestial body, moves at its own velocity.
Attractive force of the Sun
natural speed of the Earth
Earth's trajectory
This velocity would move the Earth away from the Sun were it not for the force created by the Sun's mass that pulls the planet back.
The balance between these two competing forces causes the Earth to revolve around the Sun at a fixed distance.

Newton's contribution amounted to a true scientific revolution, unifying two fields that had been regarded as completely separate.
GRAVITATION
Down with bodies!
Stars, take action!
That's actually the whole point of science–to arrive at one ultimate theory.
And that's why Newton can be considered one of the first physicists!
BEEP! BEEP! BEEP! BEEP!
Huh?
Oh, it's time for a coffee break!
I can hardly keep up. I'm not a young man anymore!
Just in time!
You seem to be doing ok ...
Who knew running from fans could be such good exercise?!
RRRRIING!
Indeed!
Speaking of youth, it's my niece!
Hello?
Oh hi, it's Gabrielle.
Hey tootsie pop! Good timing, I'm on break!
DO NOT CALL ME THAT WITH PEOPLE AROUND, IT'S SOOOO EMBARRASSING!!!
Aah, not that I'm into stars or whatever, but I was wondering if your Newton dude came up with all that because an apple fell on his head?!
Maybe I'd be better off finding a tree to sleep under than going to class.

That's my Gabby!
Sorry, but it's far from a coincidence! We can also thank him for advances in:
Theology
Philosophy
Mathematics
Astronomy
Chemistry
So he was a major geek!
And the whole apple story is probably just a myth ...
That said ...
... it's not like he was destined to become a science whiz!
GRLBRLGL
He was born prematurely, and it's truly a miracle he even survived.
I gave him a little push ... this whole humanity thing was lacking momentum ...
Then, he turned out to be a bit of a dud in school!
It was his unhealthy need to be better than everyone else that finally made him buckle down.
MWA HA HA
Oh, I see, so I'm just a little bit less of a freak than him!
Finally, when his mother insisted he take over the family farm, a few people who were aware of his talent objected and allowed him to go to university.
?
And humanity is the better for it!

Meh ...
You're making this guy out to be a real revolutionary ... yes, he had long hair, but a revolution means there's sh... chaos going down everywhere, doesn't it?
Outside Great Britain, Newton's discoveries were met with considerable hostility and were slow to catch on! Ironically, it was a man of letters (and not science) who would help spread his ideas–Voltaire!
VOLTAIRE
Newton's importance to the history of science can be summed up in an inscription supposedly written for his tomb ...
"Nature and nature's laws lay hid in night: God said, 'Let Newton be!' and all was light."
Alexander Pope, 1727
SUPER classy!
Not bad, eh?
... Oh, I gotta run. Talk later!
Nice kid, your niece!
Oops, we're running late too!
Where were we?
Two hundred years after Newton's great discoveries, another physicist would mount a serious challenge.

This one was not convinced that attractive forces could be "instant messages" between two objects, as Newton had suggested.
That would mean that anything in the vicinity of a star that suddenly appears would **instantly** feel its presence.
WHOOSH!
HELL-LLOO, I'M HEEE-RE!
Got the coffee on?
Is it ok if I smoke?
You won't even know I'm here, haha!
The young physicist believed that forces indeed acted at very high speeds, but not instantly.
So, it was his turn to spark a revolution and totally transform how we looked at gravity.
Until the early 20th century, it was thought that the space through which objects moved was itself unchanging and unchangeable. Space was seen as a stage where actors moved but which itself was outside the action.
He succeeded in showing that space and time are entities that can themselves change and be changed by other things.
He introduced a new actor, or rather a new stage: **SPACE-TIME.**
This concept would be the basis for his famous theory of **GENERAL RELATIVITY.**

You've probably already guessed this revolutionary physicist is none other than Albert Einstein!
This new theory was a game changer for gravity–it was no longer seen as an attractive force between two objects ...
... but simply as the effect of the distortion* of spacetime by an object as felt by another object.
Here's Albert the expert to explain it all to us!
Thank you ...
Ahem
Just imagine that the Universe is a large beach towel spread out on the sand.
?!
Marbles of various sizes–let's say these are celestial bodies–are scattered on the towel, which they then distort based on their mass.
*The proper term used in physics is the "curvature" of space-time

If you throw a smaller marble on this distorted towel ...
51
... its trajectory will be influenced by the impressions in the towel ...
... and it will probably end up falling into one of those impressions.
The same logic applies to a celestial body making its way through the stars.
Will you roll or shoot?
Psst, Dr. Mayor, look!
Ever seen Einstein in his un...
What's going on?
No idea
HAHAHAHAHAHA HAHAHAHAHA
Hmm ...
Ok ...
Where were we?
Nice undies, by the way ...
Even if an object's local movement always goes in a straight line, its overall trajectory can be circular.
Take a car, for example, driving in a straight line on Earth.
Locally, it keeps moving in a straight line ...
This looks very familiar...
... but overall the car will end up circling the globe
Everything truly is relative ...
In short, Einstein highlighted the dynamic nature of space-time and showed that distortions in space-time and gravity are one and the same.
Not a moment too soon!
In other words, THERE'S NO SUCH THING as gravity!

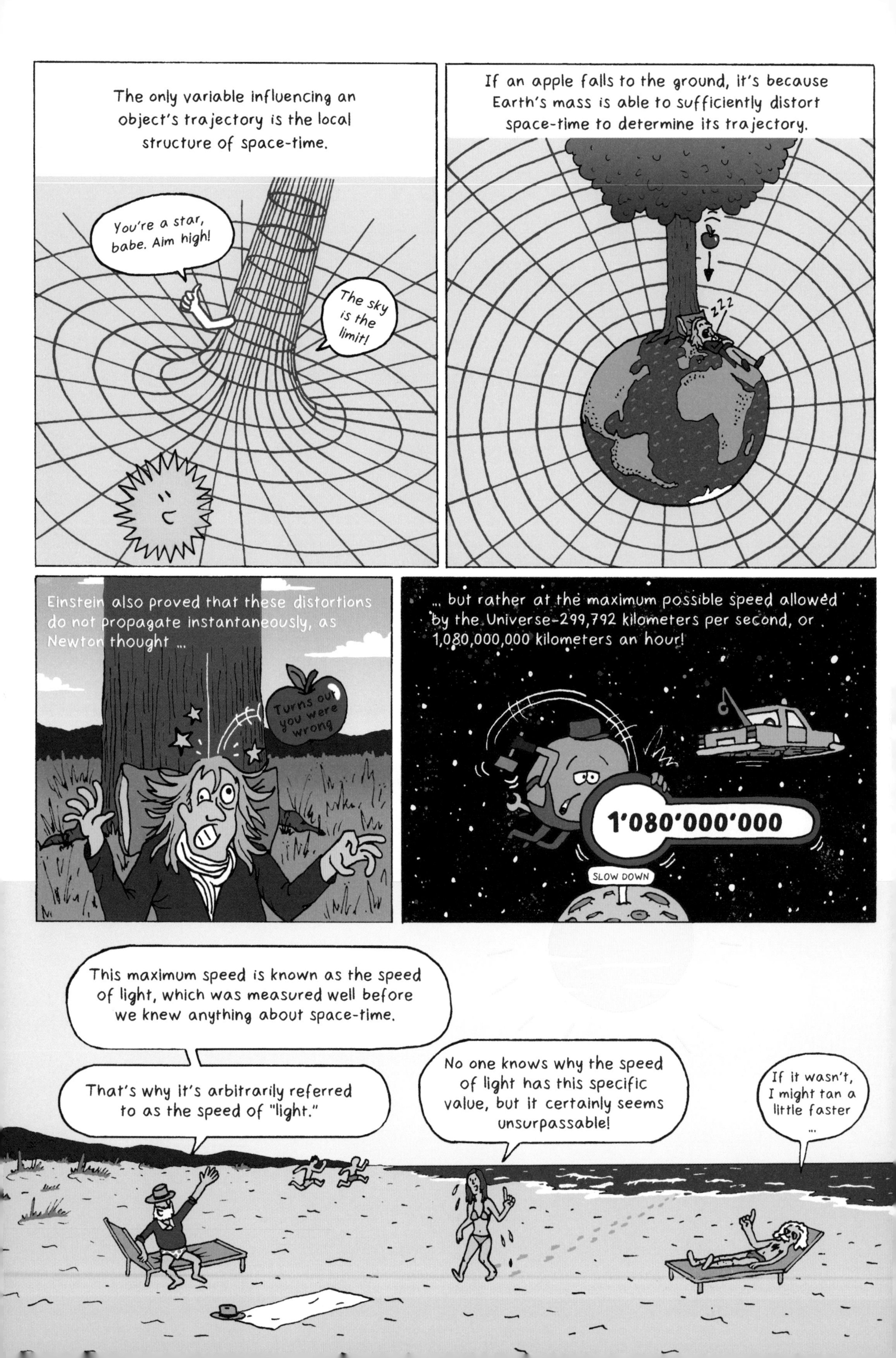

The only variable influencing an object's trajectory is the local structure of space-time.
You're a star, babe. Aim high!
The sky is the limit!
If an apple falls to the ground, it's because Earth's mass is able to sufficiently distort space-time to determine its trajectory.
Z Z Z
Einstein also proved that these distortions do not propagate instantaneously, as Newton thought ...
Turns out you were wrong
... but rather at the maximum possible speed allowed by the Universe–299,792 kilometers per second, or 1,080,000,000 kilometers an hour!
1'080'000'000
SLOW DOWN
This maximum speed is known as the speed of light, which was measured well before we knew anything about space-time.
That's why it's arbitrarily referred to as the speed of "light."
No one knows why the speed of light has this specific value, but it certainly seems unsurpassable!
If it wasn't, I might tan a little faster ...

Hold on, I just got a text from Gabrielle!
Not so busy after all ...
"In the movies, astronauts age more slowly than people on Earth. Is that BS or what?"
It's true we haven't yet looked at the "time" part of space-time.
When space-time is distorted, the distortion also relates to time because space and time are inseparable, equivalent.
Time
Space
I'd be setting you loose in hyperspace if we went into too much detail ...
So here's Hollywood to tell the story, as only Hollywood can:
BLOOD'N' GUTS!
Miller's planet
The black hole
n the movie Interstellar, the main naracter leaves the mothership for mission on "Miller's planet," located ery near a black hole 100 million mes more massive than ne Sun.
FREAKIN' BIG, MAN!
A crew member stays back on the mothership, a safe distance away from the black hole.
GRMF
GRUMF GRUMPF?
pff...
Now, we've seen that mass stretches, or "dilates," space-time – the closer we are to a mass and the larger that mass is, the more time is dilated.

If the mothership remains far enough away from the black hole for its effect to be negligible, the crew member who stayed behind would only have aged a second less than an Earthling!

Before we move on, it's important to remember we can only touch on the most significant scientific breakthroughs ...
... which means we have to jump ahead in time.
Hey Albert, you know I got a Nobel Prize too, right?
Obviously, a lot of important stuff happened between Newton and Einstein, such as Maxwell's theory of electro-magnetism, or the contributions of Poincaré and Minkowksi to mathematics.
Maxwell
Poincaré
Minkovski
Really, they still give that thing out?
Then congratulations good fellow!
All these people and many, many more added to Einstein's discoveries ...
... so it wouldn't be "fair" to give him all the credit.
Uh, yes, but ...
YES, EXACTLY!
Newton once said,
"If I have seen further, it is by standing on the shoulders of giants."
The same could be said for all famous scientists.

CHAPTER II
COSMOLOGY BIG BANG AND CMB

Let's get set to explore the darkest depths of the Universe, beginning with an explanation of cosmology.

Cosmology is the study of the Universe **as a whole**; in other words, beyond the entities we often hear about ...

Stars ...

Planets ...

Star systems ...

Galaxies ...

In a nutshell, it's studying the Universe on the largest possible scale!

That means trying to understand its origins ...

How was it "born?" How has it changed over time?

How does it work now?

Why is it still expanding?

Why are galaxies scattered the way they are, as opposed to some other way?

And what happens next?

Will the Universe ever "die?"

*Or energy.

But in order to draw more useful conclusions, we need a way to observe these distortions empirically.

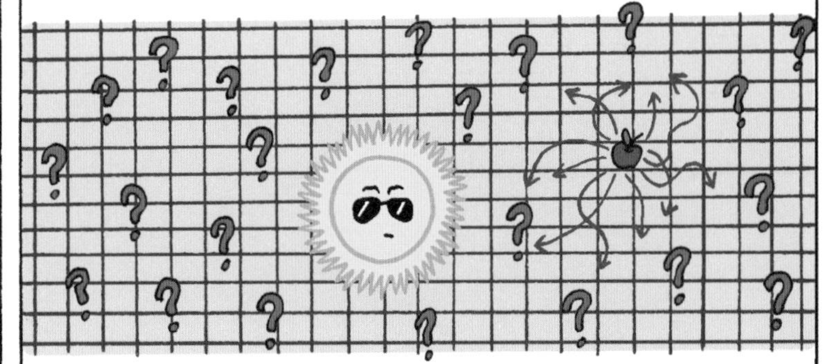

If we can calculate the distortion that happens around a celestial body, then we can predict how other objects behave when they approach it.

Now, however innovative the theory of general relativity was, it had one major flaw–Einstein's famous equations, which are the only tools we have for determining the distortion of space-time, are extremely hard to work out.

In fact, for any given situation you have to simultaneously solve a set of ten equations to calculate the "metric" that captures the distortion of space-time.

Solving Einstein's equations and determining the value of a metric for any given situation ...

... such as the area around a black hole ...

... allows us to predict the distortions of space-time that apply to that situation, regardless of time or place.

*In physics, this means finding a solution that is not exact but good enough to make predictions that are close to reality.

Since it's the presence of matter that determines the dynamics of space-time, the starting point is to define the "content" of the Universe: What type of matter is it made of and in what amounts?

According to our observations of the cosmos, the Universe is largely homogeneous, meaning that matter is distributed evenly throughout.
That might come as a surprise, because when we look up at the night sky, we see small dots of light with very large distances between them.
Even if we "zoomed out" enough for our entire solar system to be contained in a ping-pong ball ...
... the nearest solar system would be 250 meters away! That doesn't seem very homogeneous.
So the idea is to zoom out much more–until we see 100 million galaxies. At that scale, the Universe indeed starts to look homogeneous.
About 0.001 light-years
About 1,000,000 ly
About 10^9 light-years
SOLAR SYSTEM
GALAXY
100 MILLION GALAXIES

The ability to see matter as uniformly distributed throughout the Universe greatly simplified the nitty-gritty of Einstein's equations!
The four architects of the FLRW solution relied on this simplification to solve Einstein's equations and arrive at the conclusion that the Universe is **expanding.**
That means that two seemingly stationary points are constantly moving away from each other, like two ants on a balloon that's being inflated.
The fundamental difference between a balloon and the Universe is that the latter is, based on our current knowledge, infinite: it expands in nothing. It simply exists.
NOTHING
NOTHING
NOTHING
NOTHING
NOTHING
UNIVERSE
NOTHING
NOTHING
NOTHING
STILL NOTHING
Those are the key points of the Standard Model of Cosmology–
an expanding and homogeneous Universe, and a gradual dilution of the matter it contains.
This then begs the question, "What would happen if we went back in time?"
Intuitively, it makes sense that, at a certain point, the Universe was a size 0.
In other words, the famous **BIG BANG** theory!

Contrary to what it implies, the Big Bang didn't actually involve an explosion of any kind. Cosmologists prefer to use the term "Big Start."

There was no huge "matter bomb" patiently waiting to give birth to the Universe.

In fact, the Big Bang is so extraordinary that even our most sophisticated tools haven't succeeded at unlocking all its mysteries just yet.

Sooorry.

The laws of physics don't apply and even time is not a reliable concept ... in fact, it didn't even exist!

AY CARAMBA!

Even today, the Big Bang theory is just that–a theory that explains what we've been able to observe and measure.

BIG BANG

PENDING APPROVAL

It is, and may always be, a partial explanation.

We can more or less understand what happened a split second after the Big Bang ...

... but anything before that remains a mystery.

EINSTEIN'S UNIVERSE:

In the beginning, Einstein himself was not sold on the idea of a Big Bang and an expanding Universe. He believed that the Universe was static. Apparently, he later said it was the biggest mistake of his life!

BIG CRUNCH and BIG BOUNCE
Before the Big Bang, there would have been an original or primary Universe that collapsed on itself and simply bounced back out to form our Universe.
This theory now seems very unlikely and has been largely dismissed.
BIG BANG
Size
T-3 T-2 T-1 T T+1 T+2 T+3 T+4
Time
As for the future, the most credible theory is the one known as the "de Sitter model."
It describes a Universe where the expansion is accelerating.
BIG BANG
T T+5 T+10
Time
Size
Some scientists are even talking about a "BIG RIP."
Star systems will start to "rip" apart when their own internal forces are overtaken by the forces of expansion.
Wow, how badass is that?!
Of course, this is all just speculation for now. Let's get back to the Big Bang
... and make a tiny jump in time ...
... to try to get a good understanding of what happened.
The newly born Universe was nothing more than a few elementary particles of matter.
ELECTRONS ...
ELEC' 3000
... PHOTONS (light) ...
... and lots of other little things, including QUARKS.
In the beginning, these were the "ingredients" of the Universe!

They are called "elementary" particles because they are the source of all others:

Quarks formed protons and neutrons, which joined electrons to create atoms, which combined to form molecules, which formed matter as we know it.

Many different things were about to happen, so let's focus on the two most important.

The first is that quarks started to come together in groups of three, creating unbreakable protons and neutrons ...
... with each proton then "capturing" an electron, causing it to orbit that proton.
Together, they formed the very first instances of the most basic atom: **Hydrogen.**
And in this primordial soup, this process played out every time a proton encountered an electron.
The problem is that these hydrogen atoms were very shortlived due to the presence of photons, which banged into them, causing them to break apart. Those that did form survived for a very short time.
0.00001 seconds–a new world record!
Protons and electrons were thus condemned to solitary evolution, with no ability to come together permanently.
Keep in mind that, at this point, barely a second had passed since the Universe was born!
And already it measured more than **100 billion** kilometers across
... with a temperature of **10 billion** degrees Celsius.

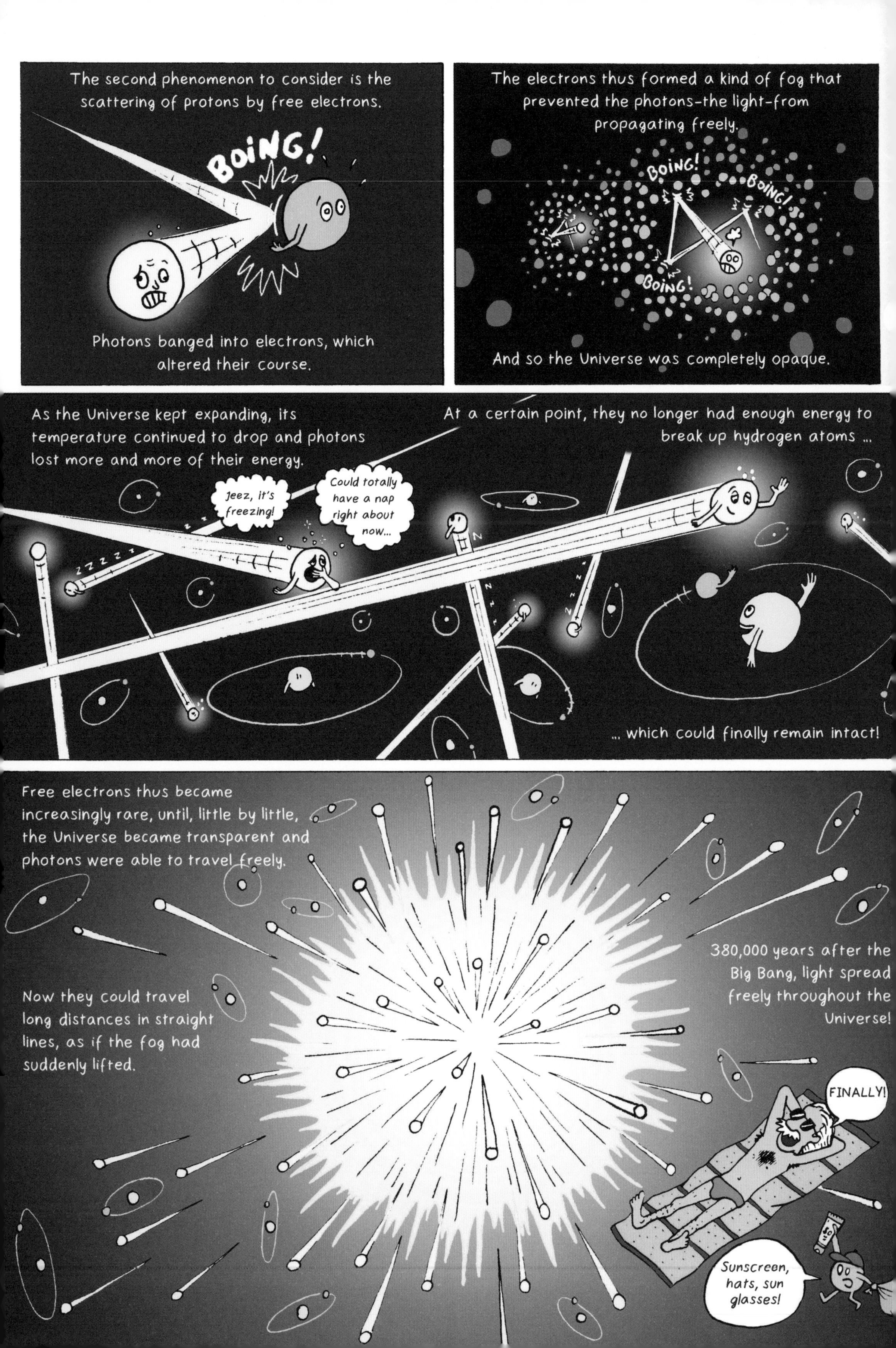
The second phenomenon to consider is the scattering of protons by free electrons.
BOING!
Photons banged into electrons, which altered their course.
The electrons thus formed a kind of fog that prevented the photons–the light–from propagating freely.
BOING!
BOING!
BOING!
And so the Universe was completely opaque.
As the Universe kept expanding, its temperature continued to drop and photons lost more and more of their energy.
Jeez, it's freezing!
Could totally have a nap right about now...
At a certain point, they no longer had enough energy to break up hydrogen atoms ...
... which could finally remain intact!
Free electrons thus became increasingly rare, until, little by little, the Universe became transparent and photons were able to travel freely.
Now they could travel long distances in straight lines, as if the fog had suddenly lifted.
380,000 years after the Big Bang, light spread freely throughout the Universe!
FINALLY!
Sunscreen, hats, sun glasses!

At that moment, as hydrogen clouds began to form (creating the first stars 100 million years later) ...
... photons embarked on a voyage that has now lasted 13.4 billion years.
What do you mean you FORGOT THE SANDWICHES?!?
This is the birth of **CMB**: The cosmic microwave background, the oldest light in the Universe.
CMB has been absolutely critical to our understanding of the Universe.
It's a little complicated, so allow me to explain:
Light is a type of what we call **electromagnetic waves.**
Light is in the same family as UV, radio waves and microwaves.*
GAMMARADIO FAMILY
Each wave has a wavelength that is unique to it and that determines what category of wave it belongs to.
Radio waves
1Km
1m
Microwaves
1mm
Infrared
1µm
VISIBLE LIGHT
UV rays
1nm
X-rays
Gamma rays
1pm
The truth is that the shorter a wave is, the greater its energy.
For example, the photons that make up UV rays have so much energy they can burn our skin.
Compare that with microwave photons, which have up to 100 billion times less energy.
*Waves and photons are one and the same! Incredible, but true!

Simply put, because the Universe is expanding, photons are constantly being stretched.

As we saw with Einstein, a body is influenced by the distortion of the space-time it is traveling in.

In fact, each photon must "pay the price" for its journey across the Universe: Very slowly, they get longer and lose energy.

The technical word for this process is "redshift." It comes from the fact that we see photons on a scale of blue (greatest energy) to red (least energy).

REDSHIFT

That feels SOOOOO good!

Huh?

Blue and full of energy, a photon crossing the Universe will get longer, losing energy in the process and becoming redder and redder; in other words, it is being "redshifted."

The FLRW Standard Model theorized about the existence of CMB.

Original CMB

CMB today

Based on our knowledge of when the CMB freed itself from the fog and its initial energy levels ...

... we can estimate the amount of energy it should have today after being redshifted for the last 13.4 billion years.

However, this prediction flows directly from measuring the redshift.

*Image : ESA - Planck (2013)

Because of this, we could finally be sure the FLRW solution was correct and the Universe was indeed expanding.

We can also prove the Universe is expanding by ... observing stars.

In fact, by determining the atomic composition of a star, we can also deduce the type of photons it emits.

ORIGINAL PHOTONS

redshift

PHOTONS RECEIVED :

Then we can compare the length of the photons received from that star with their length at the outset and estimate the magnitude of the redshift they have undergone.

Finally, by calculating the star's distance, we are able to see that redshifting increases with distance; again, proof the Universe is expanding.

Using the same approach, we can also determine the speed at which the Universe is expanding.

For instance, the expansion of the Universe would increase our distance from the Andromeda Galaxy, our neighbor 2.5 million light-years away, at a speed of about 60 km per second.

The farther away objects are, the greater their speed of expansion.

But, like, how do you even begin to measure the distance between galaxies, that are, like, super far from Earth?
That's a question for Dr. Mayor!
One common method is to look at stars known as "Cepheid Variables" that "pulsate"–their luminosity, or brightness, changes regularly.
To understand how that works, imagine a light source with an intrinsic brightness we're familiar with.
Streetlights, for example.
When we see a streetlight in the distance, we can intuitively guess how far away it is because we know the approximate brightness of a streetlight.
The same goes for a star: In order to calculate its distance, we must know both its apparent AND intrinsic brightness.

In the early 20th century, Henrietta Swan Leavitt discovered a Cepheid star's intrinsic brightness could be determined based on the speed of its pulsations.
She then showed that a star's distance could be estimated by comparing its intrinsic brightness and its apparent brightness.

Thanks to Leavitt, we could accurately calculate distances in the 10-million-light-year range ...
... in other words, distances of intergalactic proportions!

Leavitt's work also paved the way for the calculation methods used today, which can detect galaxies more than ... 30 billion light-years away!
Henrietta Swan Leavitt
Well, all hail her Majesty, Henrietta Swan Leavitt!
I have a question!
With all these discoveries, CMB and stuff like that, does this mean we know exactly everything there is in the Universe?!
No, and CMB is merely an electromagnetic snapshot!
In it, we only see primordial photons.

OK, but don't we, like, know how many planets there are in the Universe?
Hmm ... let's just say we can estimate how many there are.
Let's go in order ...
There are an estimated 2,000 billion galaxies
... in the **observable** Universe.
2,000,000,000,000 GALAXIES
Each of these galaxies has between 100 and 500 billion stars ...
So we estimate there are **at least** 200,000 billion, billion stars.
200,000 000,000,000 000,000,000
×
And we know stars have, on average, about ten planets in their orbit ... well, I'll let you do the math!
Totally, and these estimates are ten to 20 times greater than they were in 2015! Coming up with figures when dealing with such huge orders of magnitude is very complex work ...
WoOOAAHH mind-blowing!
??

At any rate, remember we are only talking about the **observable** Universe here.
We don't know if the Universe is finite or not. If it's infinite, then the numbers probably are too!
But unfortunately, time isn't, so back to work!
Uh, like, what's the next topic?
We saved the best for last ...
The MONSTERS of the cosmos!
The VIPs of sci-fi!
The biggest rock stars of the Universe ...
BLACK HOLES!
yaaaay!
CHAPTER 3
So, what exactly is a black hole?
Technically, it's the answer to Einstein's equations–the "Schwarzschild solution" mentioned earlier.
owiiiiiii
I DID IT!
In 1916, just after Einstein's theory of general relativity was published, Karl Schwarzschild found the first exact solution to those equations ...

... while serving on the frontlines with the German army during World War I!
I GOT IT!
He would die just a year later ... from an extremely rare skin condition!
What a shame!
In a nutshell ...
... take a star with a mass large enough to significantly distort spacetime.
The Schwarzschild solution succeeded in predicting the exact distortion of spacetime caused by that star.*
I emphasize **exact**–this is one of those rare cases where the solution is mathematically exact.
= X
*As well as changes in the trajectory of objects close by.
These predictions were validated a few years after Schwarzschild completed his work.
Ahhh!
Not a minute too soon!
But for present purposes, the Schwarzschild solution is relevant for another useful reason:
If we know an object's mass, we can compute its **"Schwarzschild radius."**
This term describes the radius inside which a star is so dense that nothing can escape its gravitational field.
This was just the "kickstart" Einstein's theory needed, and he became internationally famous!

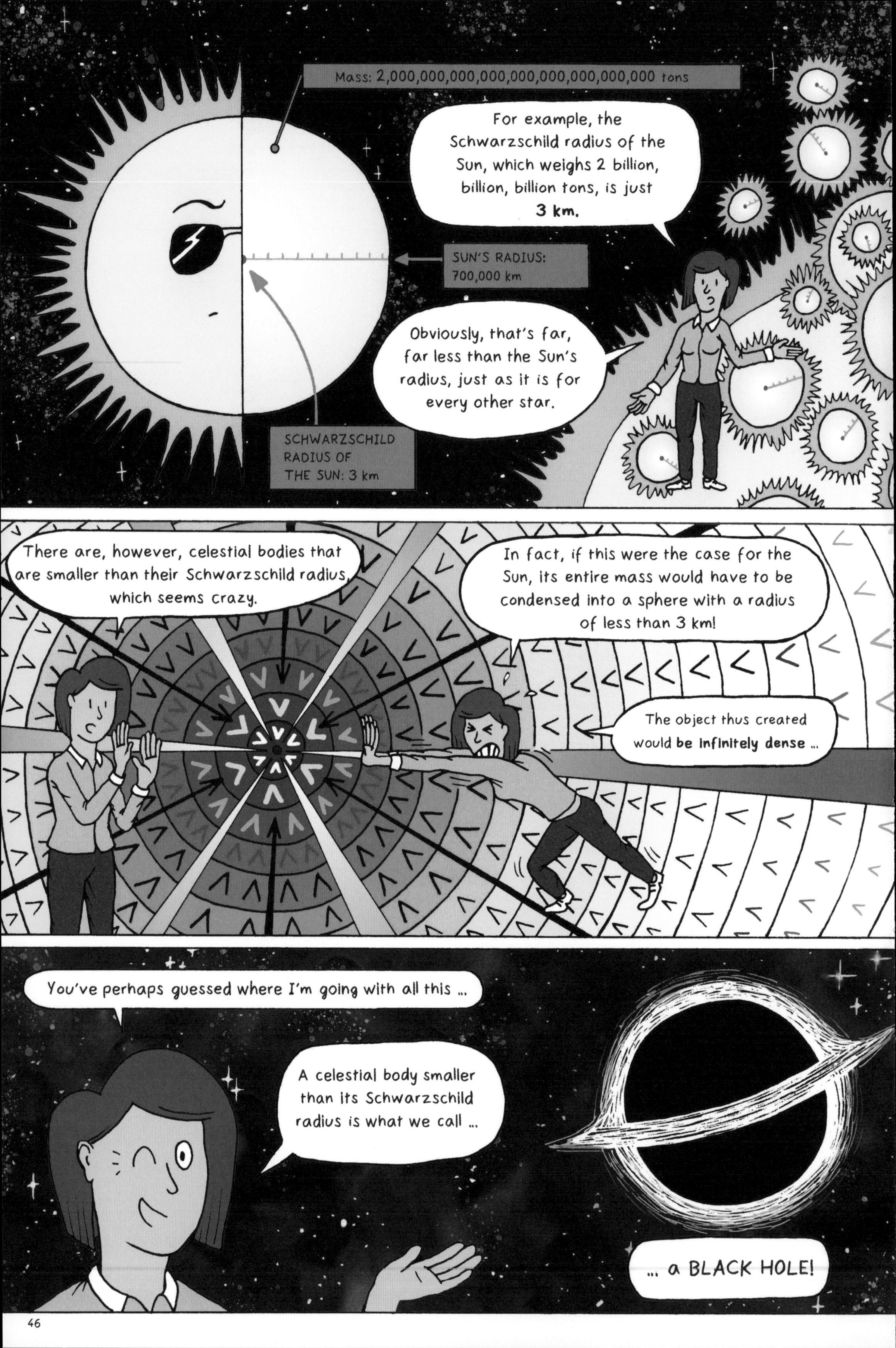
Mass: 2,000,000,000,000,000,000,000,000,000,000 tons
For example, the Schwarzschild radius of the Sun, which weighs 2 billion, billion, billion tons, is just **3 km.**
SUN'S RADIUS: 700,000 km
Obviously, that's far, far less than the Sun's radius, just as it is for every other star.
SCHWARZSCHILD RADIUS OF THE SUN: 3 km
There are, however, celestial bodies that are smaller than their Schwarzschild radius, which seems crazy.
In fact, if this were the case for the Sun, its entire mass would have to be condensed into a sphere with a radius of less than 3 km!
The object thus created would **be infinitely dense** ...
You've perhaps guessed where I'm going with all this ...
A celestial body smaller than its Schwarzschild radius is what we call ...
... a BLACK HOLE!

In addition to defining a black hole, the Schwarzschild radius also defines an invisible boundary known as the **"EVENT HORIZON"** of a black hole.

Once this boundary is crossed, it is impossible to go back!

If we are just 1 cm outside the boundary, we can, in theory, still escape ...

But if we go 1 cm inside the boundary ...

... it's game over.

You cross this boundary without ever even knowing it, without feeling anything in particular, but your fate is already sealed ...

Which is why we call it a BLACK hole–no particle of light that crosses the event horizon can escape!

For this very reason, we know very little about black holes and, basically, no one really knows what they truly look like.

Or what happens inside them.

When we look at a black hole, what we are really seeing is its event horizon.

As for its luminous disk, it's just burning matter spinning around the black hole at crazy speeds before being swallowed up.

So what would happen if you fell into a black hole??

There's really no way of knowing.

Because of the extreme distortion of space-time caused by the density of the black hole, our perception of times and distances would probably be completely turned upside down.

We might see the future of the Universe play out faster and faster before our very eyes ...

... while, from the outside, our outline would seem to be moving away in slow motion ...

... before slowly disappearing.

What happens next is hypothetical.

Death is surely inevitable, but no one knows how quickly.

Maybe we'd die instantly, as soon as we cross the event horizon ...

COME NOW!

... or we might get to "enjoy" it for a while, so to speak.

TAP TAP

Consider the giant black hole located at the center of our galaxy.*

MASS: 4 million solar masses

SCHWARZSCHILD RADIUS: 10 million km

*Most galaxies orbit around a large black hole.

In other words, they'd basically feel nothing.

But inevitably, that force would increase to 100 kg just 100,000 km from the center of the black hole.

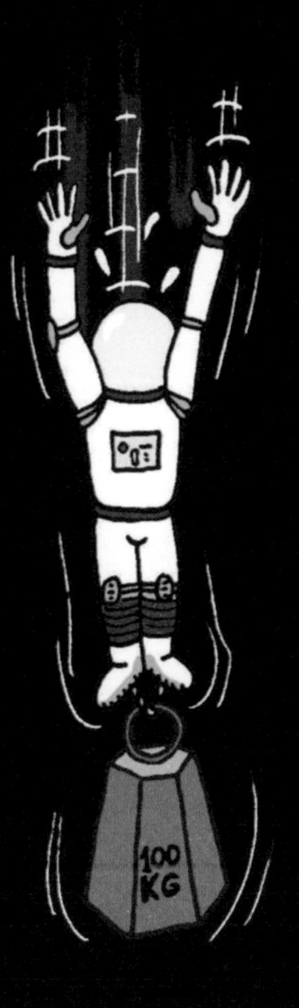

With an infinitely concentrated mass, the black hole's gravitational force would intensify so abruptly that, at some point …

Gravity = 1

… a single body would be pulled in two directions by a tremendous difference in gravity.

Gravity = 10000000000000

That's when an increasingly intensifying phenomenon called **"spaghettification"** occurs …

The body would start to disintegrate …

… more

and

more …

… until it turns into a long spaghetti noodle of matter.

And the smaller the black hole, the more sudden the spaghettification is!

AUNTIE!
Are we really going to be sucked up and turned into spaghetti?
Of course not!
Fortunately, black holes aren't cosmic vacuum cleaners!
The attractive force of a black hole is identical to that of a star with the same mass.
If, hypothetically, we replaced the Sun with a black hole of the same mass, its attractive force would stay exactly the same.
The planets in the solar system would orbit just as they do now. Of course, the temperature would take a nasty dip ...
That said, I should add that, given its already maximum density, a black hole would slowly get bigger as it absorbs other celestial bodies. A black hole has no maximum size and, in theory, it can grow infinitely ... swallowing us up in the process!
YUM YUM
CHOMP CHOMP CHOMP
The largest black hole ever observed, TON 618, weighs as much as 66 billion Suns and is capable of swallowing entire galaxies!

Conversely, according to Hawking's radiation theory (which has not been confirmed by observation), all matter within a black hole gradually evaporates. In the future, that could cause black holes to become extinct.
PHEW ...
Hot in here, isn't it?
I mean the *extremely* distant future; a tiny black hole with the same mass as the Sun would take:
10,000,000,000,000
000,000,000
000,000,000
000,000,000
000,000,000
000,000,000
billion billion billion billion BILLION
years to lose 0.0000001%
of its mass this way.
Now that we know the mind-boggling truth behind black holes, the next question is ...
What astronomical phenomenon could cause such an oddity to form in the first place?
It can happen when a very massive star-at least 20 solar masses-dies!

A star is like a huge nuclear power station where two opposite forces are engaged in a perpetual tug-of-war:

GRAVITY which would have the star collapse on itself,

and NUCLEAR REACTIONS taking place within the star* that cause it to swell.

A star's volume is determined by the equilibrium between these two forces; in fact, its very existence depends on it.

*Especially the fusion of hydrogen atoms into helium atoms.

When hydrogen, a star's main fuel source, starts to run out, the star begins to burn helium too, causing it to SWELL.

HYDROGEN

HYDROGEN + HELIUM

During this process, the star's size increases a hundredfold and most of its mass is ejected.

Eventually, the star's entire fuel supply burns out. At that point, gravity is the only remaining force ... and it takes over. The mass of the giant star implodes, and the star collapses on itself at a speed equivalent to a quarter of the speed of light ...

... rebounding with an explosion brighter than entire galaxies. That's what we call a **SUPERNOVA.**

All the remaining matter has been violently condensed into a central point. If the celestial body thus created turns out to be smaller than the Schwarzschild radius of the remaining matter, it's the point of no return. A hole has been cut in the floor of spacetime, and it's permanent!

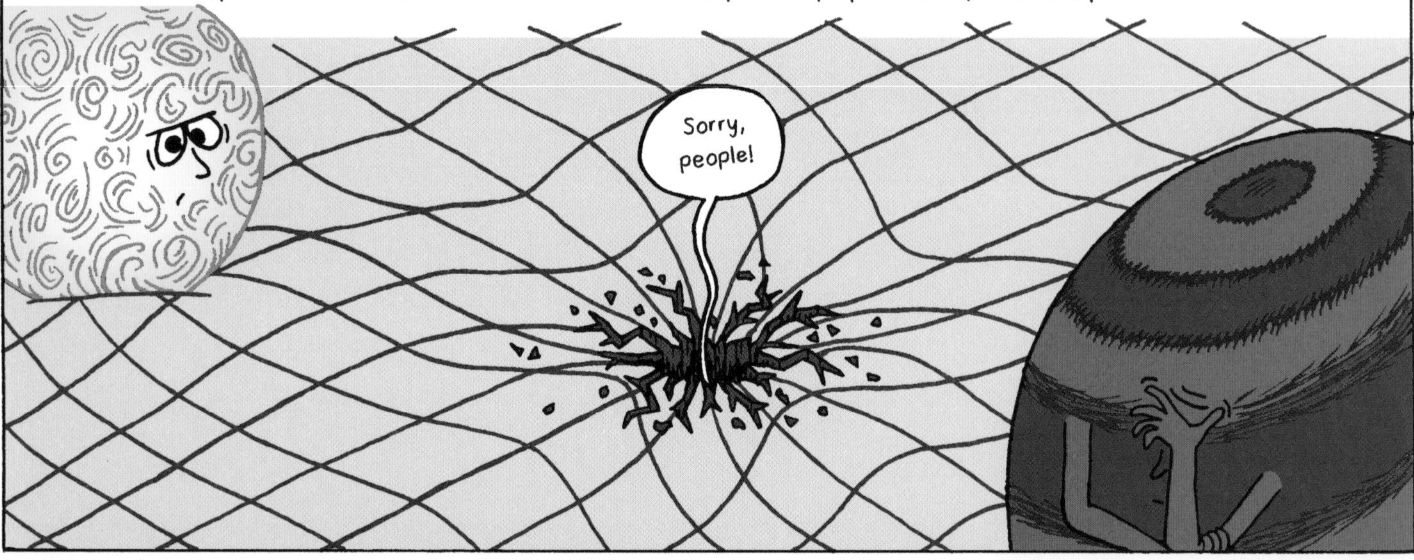

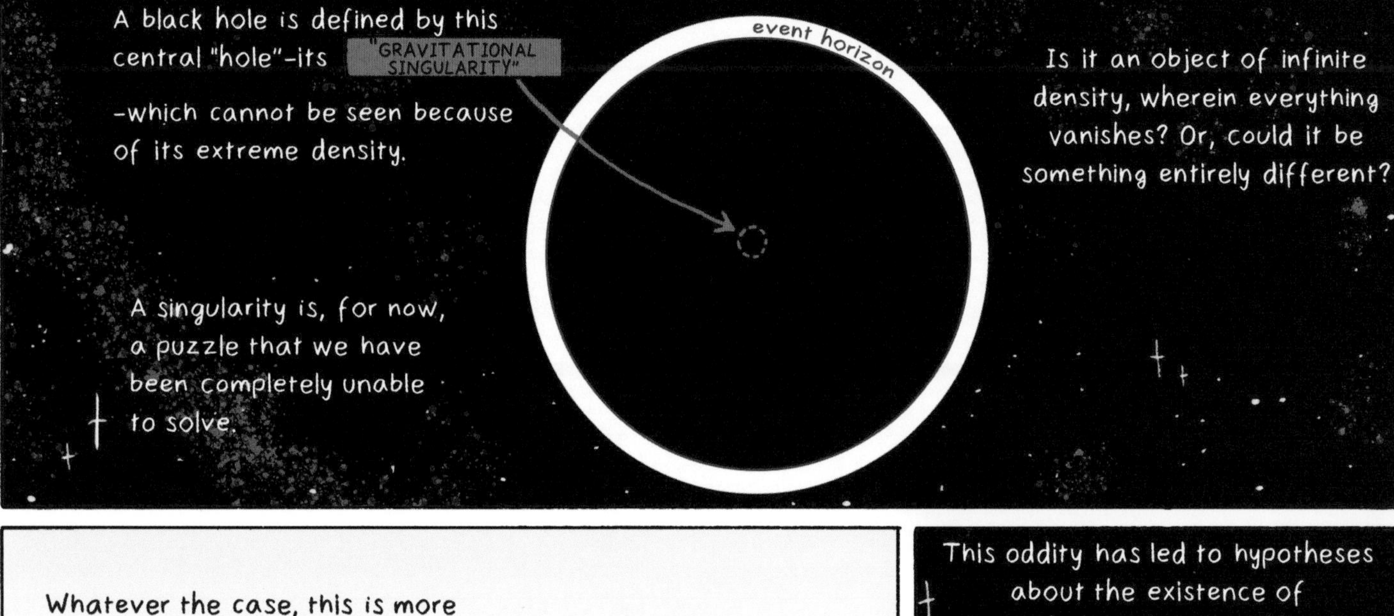

Whatever the case, this is more than a mere distortion–it is a complete *breakdown* of space-time–and it's what makes black holes so special!

BONK! BONK!

INFINITY

No matter its mass, no star can ever puncture space-time; **by definition**, that's a black hole's job.

This oddity has led to hypotheses about the existence of "wormholes" that would make intergalactic travel possible ...

DEATHSTARMAG
The OFFICIAL stellar remnant magazine for 13.7 billion years
EXCLUSIVE!
ZOMBIES OF THE UNIVERSE
SPECIAL EDITION
TYPE 1
More than 20 solar masses
Exclusive ability to die and become a
BLACK HOLE
Lifetime hype warranty![1]
[1]Expires after five years
wow effect 100% guaranteed
TYPE 2
Between 10 and 20 solar masses
Also implodes into a SUPERNOVA
AS SEEN ON TV!
Capable of transforming into a
NEUTRON STAR!
1000 in 1: 1 cm³ = 1 billion tons!
SPECIAL OFFER!
Just 10 to 30 km
in diameter!
TYPE 3
Less than ten solar masses
Turns into a
WHITE DWARF
Totally inert
100x smaller than its original star!
40x hotter than the Sun!
The fate that awaits our Sun and 97% of all stars![2]
[2]Refund subject to terms and conditions
TAY UP-TO-DATE ON ALL STAR DEATH NEWS- SUBSCRIBE TODAY!

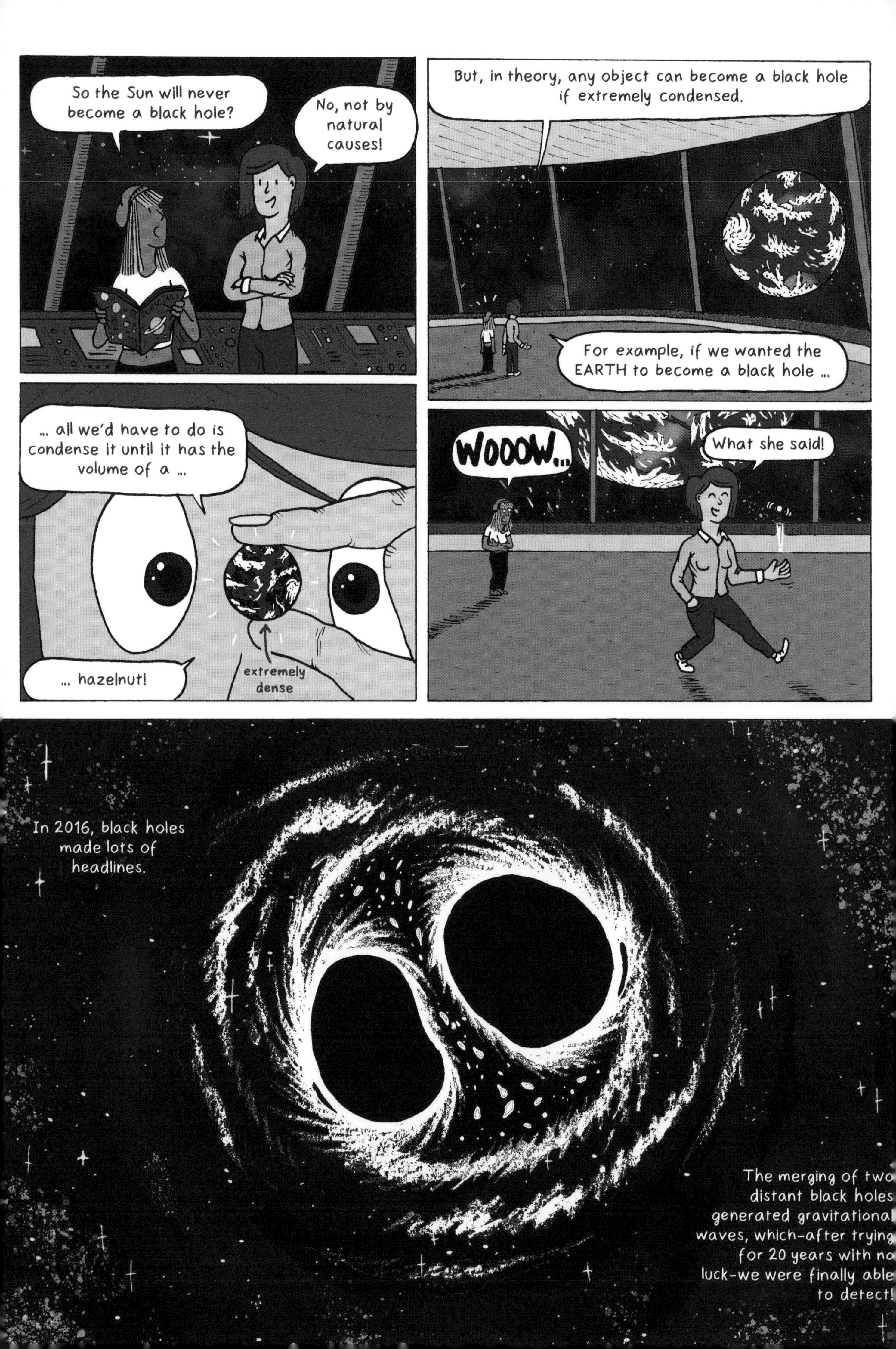
So the Sun will never become a black hole?
No, not by natural causes!
But, in theory, any object can become a black hole if extremely condensed.
For example, if we wanted the EARTH to become a black hole ...
... all we'd have to do is condense it until it has the volume of a ...
... hazelnut!
extremely dense
WOOOW...
What she said!
In 2016, black holes made lots of headlines.
The merging of two distant black holes generated gravitational waves, which—after trying for 20 years with no luck—we were finally able to detect!

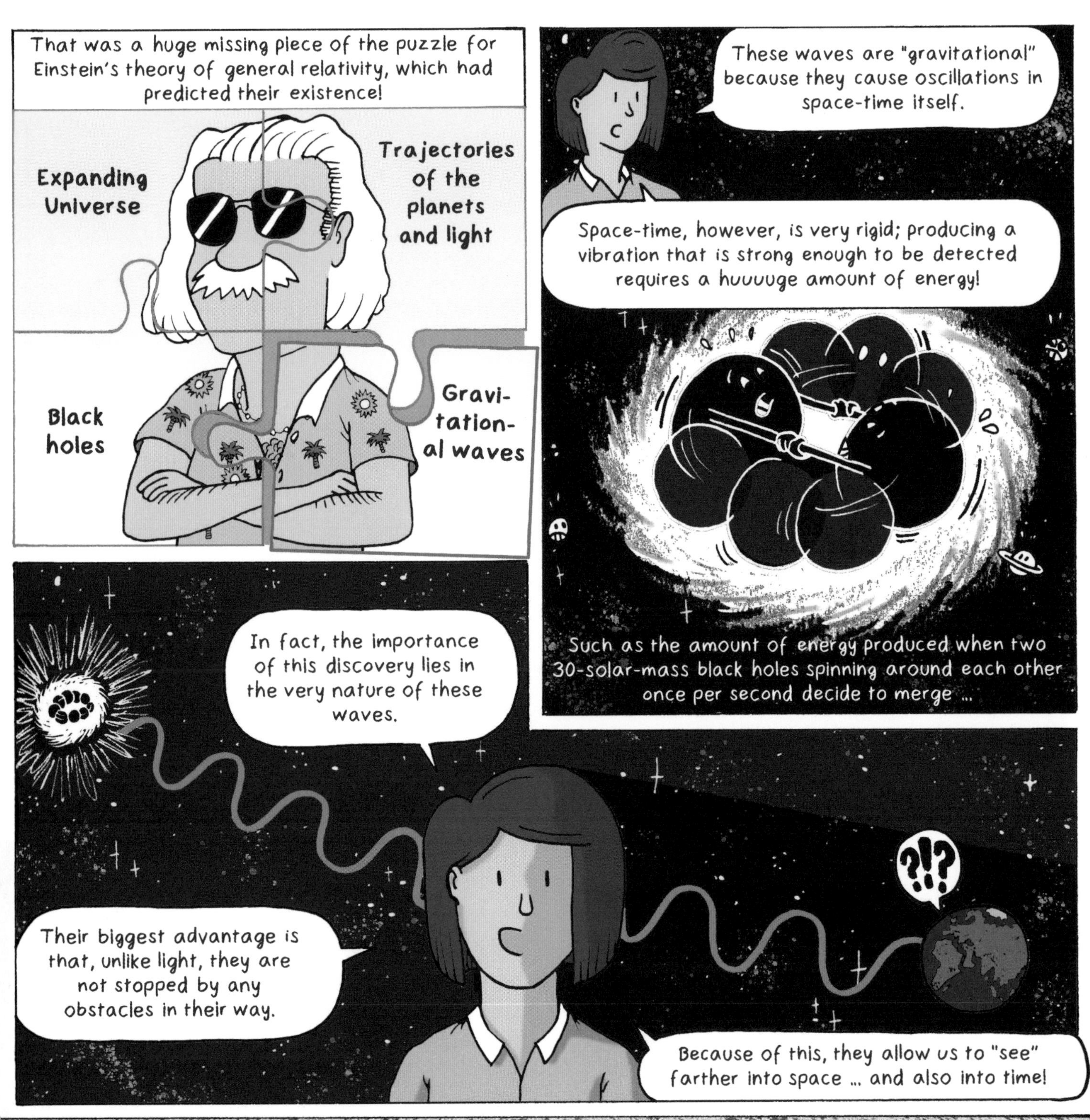

According to some theories, there should be a "CMB of gravitational waves" that would have begun its journey just a few seconds after the Big Bang. If we ever manage to detect it, it would surely provide us with a wealth of unimaginable information that could revolutionize the way we see the Universe.

And, who knows, maybe even unravel a few of its mysteries?

Noooo, it's not over already?!
Yes, I must be getting back to the lab to solve all the puzzles of the cosmos before the second volume comes out!
Hilarious ...
But hold on a sec!
If we already know everything you've explained here, then what exactly are you working on?
It's true—we cosmologists are not only working on these topics.
If we can put them in a comic book, it's because we know them pretty well already!
Apart from gravitational waves, we're also very busy looking into DARK MATTER.
The way galaxies behave leads us to believe there may well be more matter in the Universe than we have been able to observe up to now!
FLOUR
1kg
5kg
An invisible matter that, just like ordinary matter, would have a gravitational effect.

There is also DARK ENERGY, which is even more mysterious.
It alone may account for 70% of all the energy in the Universe!
~70% DARK ENERGY
~25% DARK MATTER
~5% ORDINARY MATTER
We suspect it's the reason the expansion of the Universe is speeding up when, theoretically, it should be slowing down or even reversing!
Yet, its existence still hasn't been proven.
Dark energy as a concept has only been advanced as a means of explaining the fact that expansion is accelerating.
WANTED
DEAD OR ALIVE
FOR SPEEDING UP THE EXPANSION OF THE UNIVERSE
?
POSSIBLE SUSPECTS:
1. Dark energy / 2. The Mafia / 3. Loch Ness Monster
REWARD:
NOBEL PRIZE IN PHYSICS
It so happens that dark energy and dark matter are totally misaligned with the theoretical basis of general relativity.
So, it's possible it isn't the "ultimate theory" we thought it was ...
?!?
... and we might have to rethink it to explain these strange observations.
..HHH..
..HH..
Why must we ALWAYS run?!

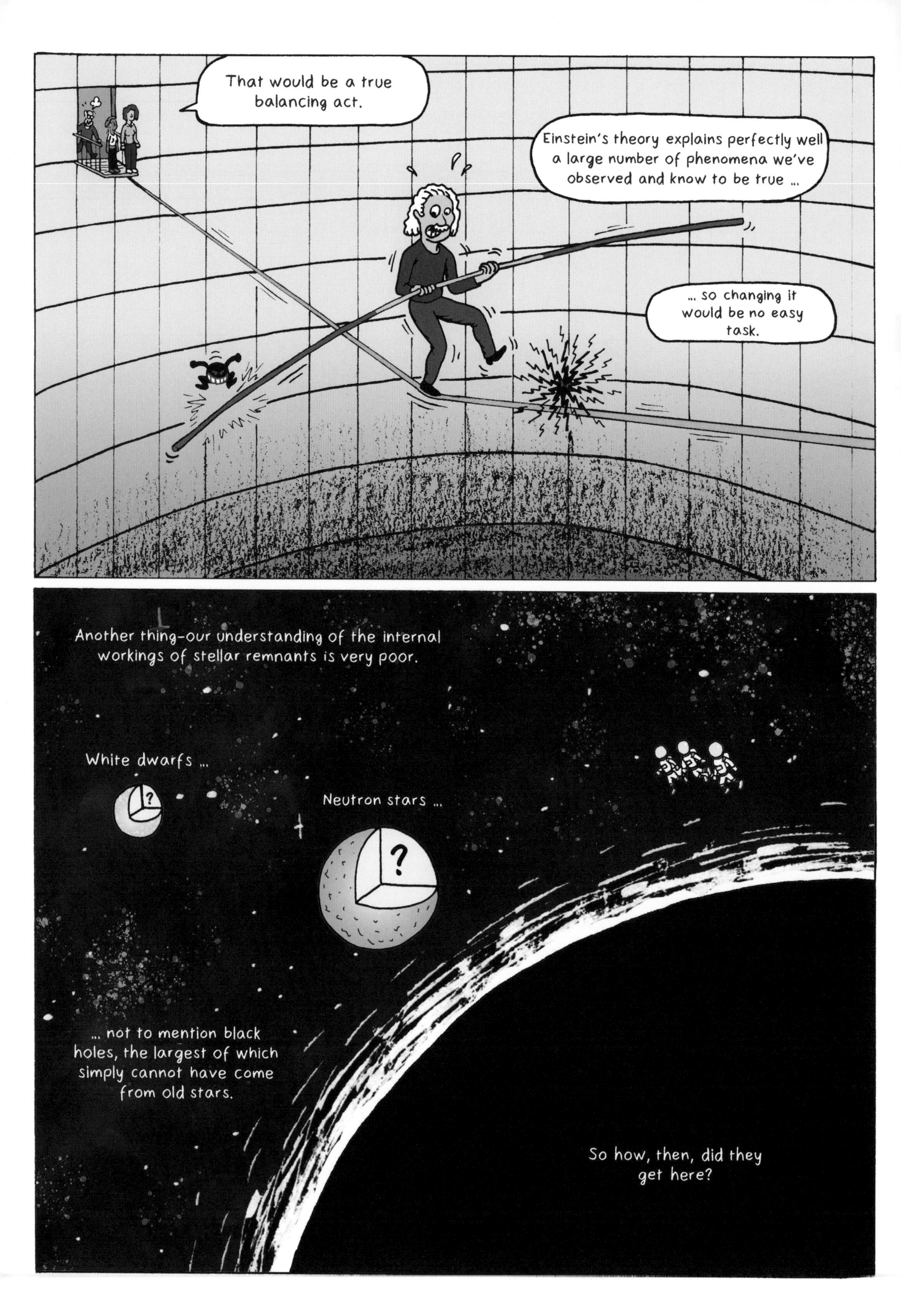
That would be a true balancing act.
Einstein's theory explains perfectly well a large number of phenomena we've observed and know to be true ...
... so changing it would be no easy task.
Another thing–our understanding of the internal workings of stellar remnants is very poor.
White dwarfs ...
Neutron stars ...
... not to mention black holes, the largest of which simply cannot have come from old stars.
So how, then, did they get here?

To conclude, I should also mention the fundamental issue of the origin of the Universe, which is still a giant question mark.

Unfortunately, we are unable to reproduce Big Bang conditions in a laboratory, which prevents us from making further inroads on the matter.

Glossary

Astrophysics
Astrophysics is the science that studies astronomical objects and their behavior—for instance, planetary systems (stars and planets) or galaxies.

Big Bang
The Big Bang is the birth of the Universe; in the FLRW model, it represents the time when the Universe was infinitely small. However, be careful—the universe wasn't just a tiny bead that exploded into something bigger. That's just how we picture it. The Big Bang is actually the moment the story of the Universe began. What happened before or a fraction of a second (something like a billionth of a billionth of a billionth of a billionth of a second) after the Big Bang remains a mystery... **See also:** FLRW.

Black Holes
Black holes are astronomical objects that are so dense they can "puncture" space-time. Every black hole possesses an event horizon. Once that horizon is crossed, there's no coming back and something called spaghettification occurs! During this process, the force pulling on our feet, for example, would be much greater than the force pulling on our head, so our body would be stretched out like a spaghetti noodle! The simplest black holes are Schwarzschild black holes, but that model is somewhat "idealistic." Kerr black holes are more realistic in that they are conceived of as black holes that rotate.

CMB
The cosmic microwave background refers to the photons that appeared during the Big Bang. For 380,000 years, they were not free to travel due to the scattering effect of free electrons. But once electrons started to bind with protons, they no longer had that scattering effect, and protons were free to travel. This is when they created CMB. CMB was sent in all directions, which is why it can be observed from Earth. Observing it (or, rather, observing fluctuations in photon energy in relation to their average energy) has provided us with a great deal of information, as well as pretty good evidence in support of the Big Bang model. **See also:** Photons, electrons, protons, FRLW, redshift.

Cosmology
Cosmology is the study of the Universe as a whole. Its origins, history, what it's made of, and what will happen to it are just some of things that intrigue cosmologists.

Curvature
Curvature is the technical term for mathematically quantifying the "distortion" of space-time. According to the theory of general relativity, curvature is a result of the presence of matter and energy. Under general relativity, the curvature of space-time is what causes objects (planets, for example) to move as they do. Therefore, there is no "gravitational force," just "deviated" trajectories caused by curvature. **See also:** Einstein's equations, space-time.

Einstein's equations
Einstein's equations are ESSENTIAL tools in the theory of general relativity. The general idea is that space-time is a dynamic entity that can be distorted (the technical term is curvature). Einstein's equations tell us that curvature is due to the presence of matter and energy. There is no single solution to these equations, but rather as many solutions as there are conceivable "situations." Typically, if we know the type of matter and energy present in any given situation, we should be able to determine the curvature throughout that situation. "Should" is the operative word, as these equations (there are ten in total) are VERY difficult to solve, and usually we have to settle for approximations. **See also:** General relativity.

Electromagnetic waves
Electromagnetic waves are special waves that travel at the speed of light. These waves come in different types, depending on their energy. From highest energy to lowest, they are: Gamma rays, X-rays, UV rays, visible light, infrared, microwaves, and radio waves. Every wave has a wavelength that roughly corresponds to its "size." A shorter wavelength corresponds to a higher-energy wave. Quantum mechanics teaches us that electromagnetic waves are, in fact, also particles—photons. This is the "wave-particle duality relation"—depending on the situation, we see waves or particles. **See also:** Visible light, photons.

Electrons
Electrons are elementary particles of the Universe that appeared during the Big Bang. Protons and neutrons make up the nuclei of atoms, around which electrons "orbit" to create complete atoms.

FLRW
The FLRW (Friedmann-Lemaître-Robertson-Walker) model, also known as the FLRW Universe, is an exact solution to Einstein's equations. In physical terms, it describes a Universe that is homogeneous (where matter and energy are evenly distributed) and expanding. It's the standard model of cosmology on which all research is based. The FLRW model predicted the existence of CMB, which is now observable. Furthermore, in an FLRW Universe, traveling light is subject to redshift—the wavelength of an electromagnetic wave traveling through the Universe will increase, while its energy will decrease. This means that light from distant stars, which has naturally had to travel for a longer time, will undergo greater redshift. And this is indeed what we see happening. These two experimental observations are, along with others, solid proof of the validity of the FLRW model. **See also:** Big Bang.